Pi is not alone!

Email:pi.is.not.alone@gmail.com

Ebook ISBN: 978-93-5311-880-8

Paperback ISBN: 9789353212131

1st publishing:2018

Published by:
Sujit kumar singh
Anuvikas township,
Boisar,
Maharashtra-401504
India

Acknowledgment

I would like to thank my wife Sweta for inspiring me in writing the book. I also thank my wonderful children Shaurya and Suryansh, for always making me smile and for understanding on those evenings when I was writing this book instead of playing games with them. I hope that one day they will read this book and understand why I spent so much time in front of my computer. I'd like to thank my parents for allowing me to follow my ambitions throughout my carrier. I dedicate this book to my father, Late R.B.Prasad, who made mathematics my first crush since my childhood. All the family members are thankful for their supports.

I would also like to thank all my seniors, juniors and friends at my workplace for their continual support.

And will not like to miss search engine GOOGLE for its support.

Contents

Introduction	4
Pi day	6
Brief history of π	8
Why π is special?	16
The search begins	18
My first step – Square	20
A Try with Triangle	25
Take a break	29
Carry on with Hexagon	30
Vehicle for Onward Journey	33
The Express Lane	39
End of the journey	43
Accuracy of π	55
Friends of π	59

Introduction

In the history of Mathematics, the most talked about number is none other than the π. Every school going knows π and none of the adults have forgotten it.

π is the sixteenth letter of Greek alphabet, transliterated in English language as 'pi' and pronounced as - /paɪ/.

The value of Pi is approximated by many; starting from the earliest civilizations – the Babylonians, ancient Egyptians, Sumerians, Chinese, Greeks and Indians. Certainly they all obtained the value empirically. The Greek mathematician Archimedes (around 3rd Century BC) is usually credited with the first rigorous Geometrical method that can approximate the value of π. Many of the later efforts were computational implementations of his method.

To understand why so many people have been fascinated, if not obsessed by π, one has to realize that π crops up almost everywhere in Calculus (which is the fundamental tool in applications of Mathematics) and, of course, inevitably in Probability and Statistics (which are indispensable in the analysis of data and information).

Since 2000 BC, even before the advent of calculus in the 17th Century, both practical and theoretical mathematicians have attempted to give better and better approximations of π. Better not just in the numerical sense, but also better in a deeper mathematical sense. It is as if discovering the exact value of π will lead to the unraveling of mysteries in some other areas of mathematics. For some mystical reason, the occurrence of π in a mathematical formula will not fail to evoke a feeling of awe and wonder in a mathematician.

Let us try to look into some other dimensions related to this mysterious Mathematical constant.

CHAPTER ONE

Pi day

Morning of 14 March 2018, the day started as usual, I logged on to my PC for daily News. Like many of us I also have Google as my homepage. The homepage was displaying a doodle – 'Pi-day'. I love this feature of Google as it keeps on surprising me with information on famous artists, pioneers, and scientists.

I was acquainted with Pi (π) (obviously we all do!), but never heard of anything like "Pi day". What is more to know about π other than the fact that its value is 3.14 and is used in relation with area and circumference of circular shapes? With the above inquisitiveness — I clicked on the doodle. It led me to the webpage www.piday.org. The website was interesting; amazed to know that people are celebrating 'Pi day' as some special occasion. My interest in Pi started growing again. Don't know why, but it was attracting me — like a magnet!

The day passed on but somewhere in my mind π has made a place – Small, but permanent place!

CHAPTER TWO

Brief history of π

18 march 2018, Sunday, the most awaited day of the week. While my family members used to wake up late on Sundays, I prefer waking up a bit early than other days. It also has a reason, may seems silly for many but for me it works fine – 'waking up early on holidays makes the holiday longer'.

π was still in my mind, knocking quietly. The question in the mind was a simple one —

"Why π is so special?"

To find the answer, as usual, I knocked the door of my dearest neighbor – Google! – First stop to solution of any problem. I searched on web and found so much information but only to know that π has a very long history and even longer digits. Pi has been known for almost 4000 years and the significant digits in π are increasing day by day.

π by Measurement

The ancient Babylonians, in about 2000 BC, calculated the area of a circle by taking 3 times the square of its radius, which gave a value $\pi = 3$.

At the same time Egyptians assumed a circle with diameter nine has the same area as a square of side eight, which implies $\pi = 3.1605$ ('*Rhind Papyrus*' 1650 BC):

$$\pi \left(\frac{9}{2}\right)^2 = 8^2 \;\Rightarrow\; \pi = \frac{256}{81} = 3.16049$$

A Biblical verse indicates that π was approximated to 3:

"*And he made a molten sea, ten cubits from the one brim to the other: it was round all about, and his height was five cubits: and a line of thirty cubits did compass it about.*"

—I Kings 7:23 (King James Version)

These values were based on measurement of circle and not by any calculations.

It is worth mentioning that till this time π was correct only up to two significant digits. Digits after 3.1 were varying.

π by Calculation

The first calculation of π was done by Archimedes of Syracuse, around 250 BC. Archimedes approximated the area of a circle in between areas of two regular polygons:

The polygon inscribed within the circle and the polygon within which the circle was circumscribed.

The areas of polygons gave upper and lower bounds for the area of circle. Archimedes knew that he had not found the value of π but only an approximation within those limits. Archimedes showed that π is between

$$\frac{223}{71} < \pi < \frac{22}{7} \Rightarrow 3.1408 < \pi < 3.1428$$

It was first instance when π was correct up to three significant digits. π = 3.14 (the value which we are still using!).

Ptolemy, 150 AD, used some series to calculate π up to 4 digits:

$$\pi = 3 + \frac{8}{60} + \frac{30}{60^2} = 3.14167$$

Around AD 265, Chinese mathematician Liu-Hui created another polygon based method and calculated π=3.1416.

AD 480, Zu Chongzhi adopted polygon method and calculated π up to 8 digits. This record held for another 800 years.

AD 499, Aryabhata of India calculated

$$\pi = \frac{62832}{20000} = 3.1416$$

AD 499, Brahmagupta of India calculated $\pi = \sqrt{10}$ (though it is having only one digit of accuracy but easy to remember!).

AD 1424, Al Kashi, Islamic mathematician, calculated π to 14 digit accuracy. This was the highest accuracy achieved till that time.

Many years later, Ludolph van Ceulen (1610) gave an estimate that was accurate to 34 decimal places. He used Archimedes' method (based on a 262 sided polygon).The digits were later used to adorn his tombstone.

π by Analytical Methods

François Viète, French mathematician, formulated the first instance of an infinite series for calculating π. It was published in 1593 and opened new gate for calculating π. His formula, an infinite series, calculated π to an accuracy of nine decimal digits, not a great accuracy on those days but the method was new:

$$\frac{\pi}{2} = \frac{2}{\sqrt{2}} \times \frac{2}{\sqrt{2+\sqrt{2}}} \times \frac{2}{\sqrt{2+\sqrt{2+\sqrt{2}}}} \cdots$$

In the later 17th century, Newton and Leibnitz founded Calculus. This powerful tool was quickly exploited to find new formulae for π.

Gregory-Leibnitz use the below arctan series

$$\tan^{-1} x = \int_0^x \frac{dt}{1+t^2} = x - \frac{x^3}{3} + \frac{x^5}{5} - \frac{x^7}{7} \cdots$$

And by replacing x=1 the famous Gregory-Leibnitz formula is obtained

$$\frac{\pi}{4} = 1 - \frac{1}{3} + \frac{1}{5} - \frac{1}{7} + \frac{1}{9} \cdots$$

John Machin from London, in 1706, reached 100 decimal places with a fast converging arctan formula

$$\frac{\pi}{4} = 4\,arctan\,\frac{1}{5} - arctan\,\frac{1}{239}$$

The same year, William Jones (Welsh mathematician) used the symbol $\boldsymbol{\pi}$ to represent the ratio of the perimeter of a circle to its diameter. Use of the symbol was popularized by Euler, who adopted it in 1737, and soon it became a standard.

In 1761, Johann Heinrich Lambert (Swiss mathematician) proved the **irrationality** of π.

In 1873 William Shanks (British) estimated π up to 528th digits.

After the invention of desktop calculators, D. F. Ferguson (1947) raised the accuracy to 808 decimal digits.

In 1882, Carl Louis Ferdinand von Lindemann (German) proved that π is **transcendental**.

With the development of computer technology in the 1950s, π was computed accurately to thousands and then millions of digits.

The first computer calculation of Pi was performed on a device named ENIAC (Electronic Numerical Integrator and Calculator of US Army). The 1949 computation of π to 2,037 places on ENIAC took a time…

Any guess?

Not in the dreams can you imagine, It took 70 hours!

New infinite series formulas were discovered by Srinivasa Ramanujan (India) in1910, one of these series is remarkable:

$$\frac{1}{\pi} = \frac{2\sqrt{2}}{9801}\sum_{k=0}^{\infty}\frac{(4k)!\,(1103 + 26390k)}{(k!)^4 396^{4k}}$$

Each term of this series produces an additional eight correct digits in the result. Gosper used this formula to compute 17 million digits of π in 1985.

David and Gregory Chudnovsky used Ramanujan's formula with slight modifications and calculated record computation to over four billion decimal digits in 1994.

Since 1980, Yasumada Kanada is one of the major actor in the race to compute huge number of digits for π. Most of his calculations are made on supercomputers and are based on modern high order iterative algorithms.

Up to trillionth digits of π is known and the numbers will keep on increasing with the invent of powerful computers.

I would like to confirm to all readers that this is just a fraction of details about timeline of π. There may be certain other important milestones and names which I have missed.

Actually the list of persons who contributed to the topic 'π' is just like the digits in π — it is keep on increasing day by day. So it is a herculean task to compile all the details.

CHAPTER THREE

Why π is special?

I have a great belief:

'Laws made by nature are general and applies to all, only human makes laws for specifics.'

And only this belief forced me to think– why nature has created a shape, Circle, which is so different than the other shapes that a special constant 'π' only applies to it?

These are situations where Google or internet will not help you. That is the limitation – internet will show only what is already discovered and available on net, for any new idea one has to use their own mind. Here I also felt why computers can never replace human mind?

The uniqueness of π was making me uncomfortable. I was not jealous of π but I am worried that my beliefs on natural laws were at stake. And who likes ones beliefs to be broken?

Finding no reason for uniqueness of circle I changed my search the other way — there must be other shapes which have similar properties as that of circle. And finding these shapes is the only way to keep my belief on natural laws intact.

CHAPTER FOUR

The search begins

I started my expedition to find Shapes having properties similar to circle. The first thing that came into mind was 'Regular Polygons'.

The word Polygon is made from two words '**poly**' and '**gons**'. The terms 'Poly-' and '-gons' have Greek origin and there meaning is '***many***' and '***angles***' respectively. So a polygon is made of many straight lines and has many angles, examples are triangle, square, rectangle, pentagon, hexagon etc. The term 'Regular polygon' is used for those polygons which have all their sides of same length, such as: equilateral triangle, square, regular pentagon etc.

In fig.-1 we can see that if number of sides of a regular polygon is increased then the shape will resemble more like a circle.

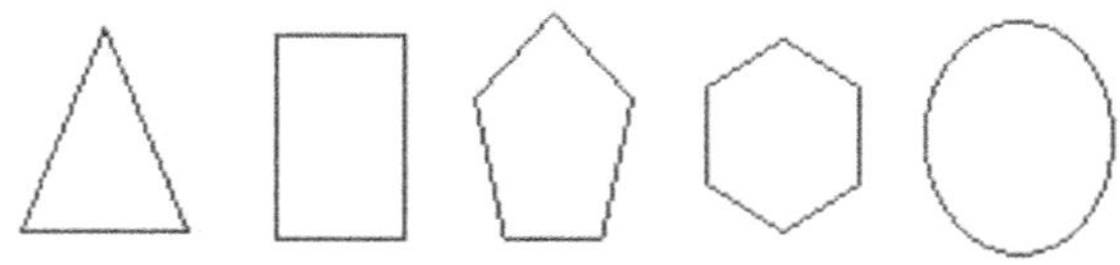

Figure-1

Assuming that answer can be found in these polygons, I started to work on a primitive question:

What is common in between an equilateral triangle, square and a circle?

Following are some special property of circle with respect to π:

1) Ratio between the perimeter of circle and its diameter is a constant (π).
2) The perimeter of circle is $2\pi r$, where r is the radius of circle.
3) Area of circle is πr^2.

Availability of similar properties in other regular polygons can help me in finding the solution of bigger problem.

CHAPTER FIVE

My first step – Square

Square is the first and simplest regular polygon that came into my mind. The reason may be its simplicity and ease of calculation.

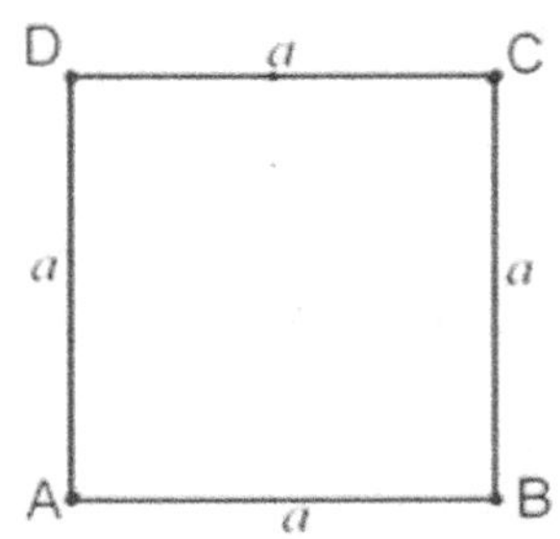

The above figure (ABCD) resembles a Square with each side having length – 'a'.

To find any similarity between square with the first property of circle –'*Ratio between the perimeter of the circle and its diameter (2* x *radius) is a constant'*, we need something in this square shape which is equivalent to the Diameter or Radius of circle.

Distance from center of the square to the corner (OD in the fig. below) was my first guess. But, after lots of calculations and experiments it did not provided any positive result.

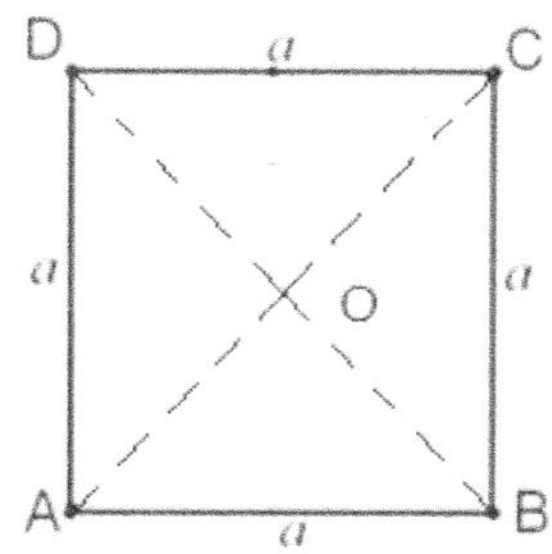

The second option was distance from center of the square to the midpoint of any one side of the square (Apothem of square).

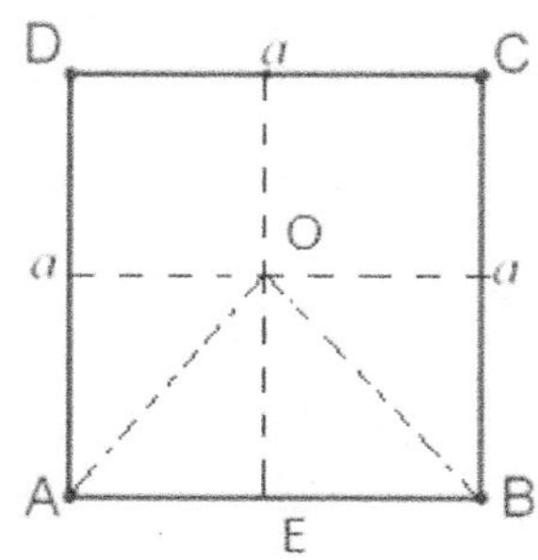

Let us assume OE is our substitute for radius of circle, we name it $\mathbf{R_4}$ (radius equivalent for '4 sided polygon' or the Apothem)

So,

$$R_4 = OE = \frac{a}{2}$$

With the available substitute we are ready to try our first check:

'Ratio between the perimeter of the circle and its diameter is a constant'

Perimeter of square, $P_4 = 4a$,

Ratio of perimeter to two times R_4 (or 2 R_4)

$$\frac{P_4}{2R_4} = \frac{4a}{2 \times \frac{a}{2}} = 4$$

It is worth noting that the ratio is independent of measurement of square and thus it is constant for all squares. The ratio for square is '4' (let's name it π_4 – a constant for 4-sided regular polygon).

The first check is a pass, now let's go for the next one :

'The perimeter of circle is 2πr, where r is the radius of circle.'

We already have assumed that R_4 and π_4 of square are radius and π equivalent for circle. Let's check what is $2\pi_4 R_4$ for a square?

$$2\pi_4 R_4 = 2 \times 4 \times \frac{a}{2} = 4a$$

This is same as Perimeter of square

So, we can say that:

'The perimeter of square is $2\pi_4 R_4$, where R_4 is Apothem of square and π_4 is a constant for square shape.'

With the second check also successful you must be enthusiastic to check the third one, me too.

Let's do it –

'Area of circle is πr^2'.

For square,

$$\pi_4 R_4^2 = 4 \times \left(\frac{a}{2}\right)^2 = a^2$$

This is same as area of square.

Wow! Area of square by this method is same as the conventional one (= side^2).

It seems we have found at least one shape which is similar in nature with the circle. It has a constant, π_4, like a circle and its value is 4 (little more than 3.14 or π).

CHAPTER SIX

A Try with Triangle

The results with Square shape poured optimism into me. But, there was a bit of fear of coincidence also. It has to be confirmed with other shapes.

I quickly moved towards another shape, the equilateral triangle, to negate the probability of coincidence. Equilateral triangle is the simplest form of regular polygon with three equal sides.

Let us assume an equilateral triangle Δ ABC with each side = a.

Point D and E are midpoints of side BC and BA. Perpendiculars are drawn at both points D and E. Both the perpendiculars intersect at O.

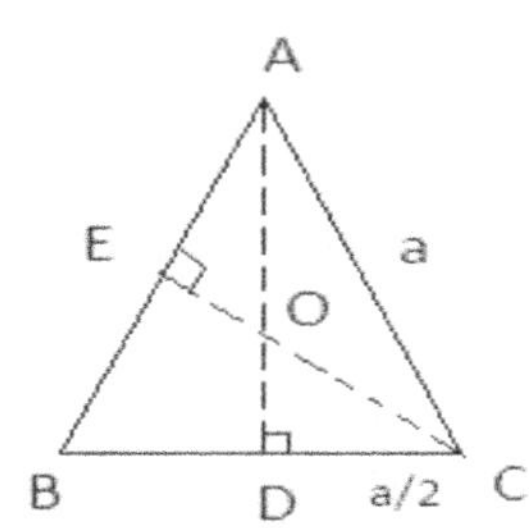

O is the center of Δ ABC from where all the sides are at minimum and equal distance.

As we know that perimeter of equilateral triangle is 3a and its area is $\sqrt{3}/4\ a^2$.

Let's check our luck with triangle.

In the ΔABC, Apothem OD will be our substitute for radius of circle and we name it R_3 (radius equivalent for a three sided polygon).

$$R_3 = OD = \frac{a}{2\sqrt{3}}$$

(OD = CD.tan30°, a little trigonometry)

<u>Our first check</u>:

Perimeter of triangle, $P_3 = 3a$,

Ratio of perimeter to two times R_3 ($2R_3$)

$$\frac{P_3}{2R_3} = \frac{3a}{2\left(\frac{a}{2\sqrt{3}}\right)} = 3\sqrt{3}$$

The result is independent of sides of triangle and thus the ratio is constant for all regular triangles and is $3\sqrt{3}$ or 5.196 (= π_3, a constant for 3 sided regular polygon).

The second check

'The perimeter of circle is 2πr where r is the radius of circle.'

For a triangle, R_3 and π_3 are radius and π equivalent of circle.

$$2\pi_3 R_3 = 2 \times 3\sqrt{3} \times \frac{a}{2\sqrt{3}} = 3a$$

Wow! That is the Perimeter of triangle.

And the final one

'Area of circle is expressed as πr^2'

For triangle,

$$\pi_3 R_3^2 = 3\sqrt{3} \times \left(\frac{a}{2\sqrt{3}}\right)^2$$

This is area of equilateral triangle.

Not less than a Jackpot!

We have found another close friend of circle, triangle (can you believe it?) Or we can say friend of π.

We can conclude with certainty:

"All equilateral triangles have a constant, π_3, *which has a value 5.196 or* $3\sqrt{3}$*."*

CHAPTER SEVEN

Take a break

Ah! Now I can take a relief.

Why not?

It cannot be a coincidence, certainly not with after knowing two shapes!

My beliefs about natural laws are till now intact.

Till now two friends of π are discovered, π_3 (3 $\sqrt{3}$) and π_4 (4).Are there other friends of π too, getting ready to be unearthed!

With two of the simplest regular polygons showing similar property as that of circle, it is certain that something is common with regular polygons and circle. We are on the right path. Let us check our belief with few more polygons.

CHAPTER EIGHT

Carry on with Hexagon

You may be wondering! Why Hexagon (6 sided polygon) and why not pentagon (5 sided polygon)?

The answer is a simple one. Formulas used for calculating area of a regular Pentagon are a bit complex and may irritate some of the readers who doesn't like math so much (but are eager to know about π). Readers can take it as an exercise to calculate area of a Pentagon. However I am quite sure that after reading the book you will have no questions left.

Let us assume a regular hexagon ABCDEF with each side of length= a.

Point H and K are midpoints of side EF and FA. Perpendiculars are drawn at both points H and K. Both the perpendiculars intersect at O.

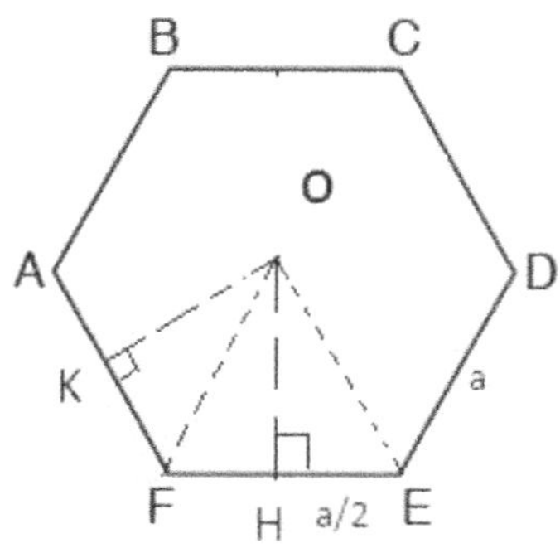

O is the center of hexagon, from where all the sides are at minimum and equal distance.

A regular hexagon can be viewed as a combination of six regular triangles. So its perimeter will be 6a and the area is 6 times ($\sqrt{3}/_4\, a^2$) or ($3\sqrt{3}/_2\, a^2$)·

Apothem OH will be our substitute for radius of circle and we name it R_6 (radius equivalent for 6 sided regular polygon)

$$R_6 = OH = EH \tan 60^0 = a\frac{\sqrt{3}}{2}$$

And the constant π_6 ,

$$\pi_6 = \frac{Perimeter}{2 \times Apothem} = \frac{P_6}{2R_6} = \frac{6a}{2\left(\frac{a\sqrt{3}}{2}\right)}$$

$$\pi_6 = 2\sqrt{3} = 3.464$$

And, area of hexagon = $\pi_6 . R_6^2$

$$= 2\sqrt{3} \times \left(\frac{a\sqrt{3}}{2}\right)^2 = \frac{6\sqrt{3}}{4} a^2$$

This is 6 times area of regular triangle.

Now there seems no doubt about existence of a relationship between regular polygon and circle and also about existence of other friends of π.

CHAPTER NINE

Vehicle for Onward Journey

As the sides of polygon increases, the calculations for finding the π equivalents for polygons will become more complex. We need a general formula that applies to all type of regular polygons.

Let's revisit our previous three shapes.

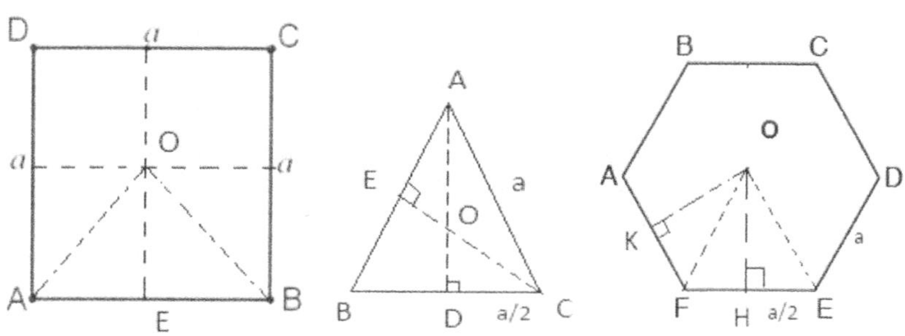

In all the three shapes, we were actually interested in finding Radius equivalent for the polygons or we can say the Apothem – the distance between center of polygon (formed by intersection of perpendicular bisectors of two adjacent sides) and the midpoint of any side. For square it was OE, for triangle it was OD and for hexagon it was OH.

And for all the cases, it was obtained with the help of the triangles formed with one side of the polygon and the center O. ΔOAB in the square, ΔOBC in the triangle and ΔOFE in hexagon.

Let us take the example of square ABCD.

Square is a Regular polygon with ‘n’ sides, for square n=4.

We all know that at any point sum of all the angles is 360^0.

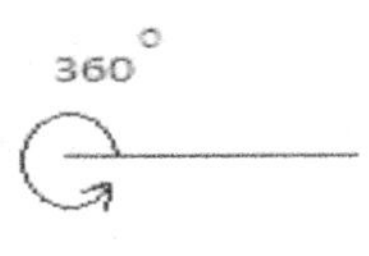

So angle at point O of square ABCD is 360^0.

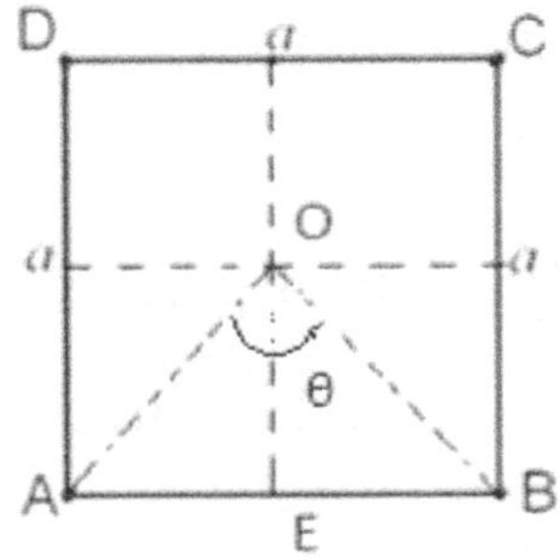

Let θ is the angle formed by AB on point O.

Since it is a regular polygon, all sides AB, BC, CD and DA will form equal angle θ at O.

$$\angle AOB = \theta = \frac{360^0}{number\ of\ sides}$$

$$\theta = \frac{360^0}{n}$$

In ΔEOB,

Since EO is perpendicular bisector of AB, Hence,

$$\angle EOB = \frac{\theta}{2} = \frac{180^{\circ}}{n}$$

And also,

$$EB = \frac{a}{2}$$

So,

$$\tan\frac{\theta}{2} = \frac{BE}{OE} \Rightarrow OE = \frac{BE}{\tan\frac{\theta}{2}}$$

$$\Rightarrow \boldsymbol{R_n} = \frac{\boldsymbol{a}}{\boldsymbol{2\tan\frac{\theta}{2}}}$$

Also, Perimeter of ‘n’ sided regular polygon, with each side ‘a’ is

$$P_n = n\,a$$

After having R_n and P_n for the polygon we can move ahead and do our calculations for regular polygon as we did earlier.

We know that:

‘Ratio of perimeter (P_n) and twice the radius equivalent (R_n) is a constant.’

So,

$$\pi_n = \frac{P_n}{2R_n} = \frac{na}{2 \times \dfrac{a}{2\tan\dfrac{180^\circ}{n}}}$$

$$\Rightarrow \boldsymbol{\pi_n = n\tan\frac{180^\circ}{n}}$$

$\boldsymbol{\pi_n}$ depends only upon n, number of sides of regular polygon. So if we decide the number of sides of a polygon then π_n becomes constant for that polygon.

With the help of this constant we can derive perimeter and area of the n sided polygon (having each side 'a'):

$$\mathbf{P_n = 2\,\pi_n R_n},$$

$$\mathbf{A_n = \pi_n\, R_n^{\ 2}}$$

We have devised the formula and now we have to validate its applicability with our known shapes. Let's do it build the confidence.

TABLE-1 shows values of π_n , $R_{n,}$, $P_{n,}$ A_n for different regular polygons using the formula. Here n is the number of sides of polygon like n=3 for triangle and 4 for square.

Table 1

n	π_n	R_n	P_n	A_n
3	$3\sqrt{3}$	$\frac{a}{2\sqrt{3}}$	3a	$\frac{\sqrt{3}}{4}\ a^2$
4	4	$\frac{a}{2}$	4a	a^2
6	$2\sqrt{3}$	$\frac{\sqrt{3}}{2}a$	5a	$\frac{6\sqrt{3}}{4}\ a^2$

The table shows that the devised formulas are valid for all known shapes and we hope for all upcoming shapes.

We have found our vehicles for onward journey. They will help us to find other friends of π and finally reach our destination, the circle.

Our vehicles are:

$$\pi_n = ntan\,\frac{180°}{n},$$

$$R_n = \frac{a}{2tan\,\frac{180°}{n}},$$

$$P_n = 2\pi_n R_n\ ,$$

$$A_n = 2\pi_n R_n^2$$

CHAPTER TEN

The Express Lane

There is no point in discussing applicability of the formulas derived in previous chapter for individual shapes. Instead we will find the areas of different known regular polygons and then compare it with the areas obtained by our method.

Again internet will help us. Areas of different regular polygons are collected from internet (Wikipedia). Areas are again calculated by our method using Excel worksheets. Values up to 5 decimal points are taken without rounding off the last digit.

In the next table we assume that the polygon is having each side as 'a' and for calculation of π_n, $\mathbf{R_n}$, $\mathbf{P_n}$ and $\mathbf{A_n}$. following formulas will be used:

$$\pi_n = \text{n}\tan\frac{180°}{\text{n}},$$

$$R_n = \frac{\text{a}}{2\tan\frac{180°}{\text{n}}},$$

$$P_n = 2\pi_n R_n,$$

$$A_n = \pi_n R_n^2$$

Table 2

Sides n	Name of regular polygon	Area from internet	R_n	π_n	P_n	A_n
3	Triangle	0.43301a^2	0.28867a	5.19615	3a	0.43301a^2
4	Square	a^2	0.5 a	4	4a	a^2
5	Pentagon	1.72047a^2	0.68819a	3.63271	5a	1.72047a^2
6	Hexagon	2.59807a^2	0.86602a	3.46410	6a	2.59807a^2
7	Heptagon	3.63391a^2	1.03826a	3.37102	7a	3.63391a^2
8	Octagon	4.82842a^2	1.20710a	3.31370	8a	4.82842a^2
9	Nonagon	6.18182a^2	1.37373a	3.27573	9a	6.18182a^2
10	Decagon	7.69420a^2	1.53884a	3.24919	10a	7.69420a^2
11	Hendecagon	9.36563a^2	1.70284a	3.22989	11a	9.36563a^2
12	Dodecagon	11.19615a^2	1.86602a	3.21539	12a	11.19615a^2
13	Tridecagon	13.18576a^2	2.02858a	3.20421	13a	13.18576a^2
14	Tetradecagon	15.33450a^2	2.19064a	3.19540	14a	15.33450a^2
15	Pentadecagon	17.64236a^2	2.35231a	3.18834	15a	17.64236a^2
20	Icosagon	31.56875a^2	3.15687a	3.16768	20a	31.56875a^2

R_n = Apothem of a regular polygon: A line segment from the center of polygon to the midpoint of one of its sides.

Areas of 21 and more sided regular polygons were not easily available on internet.so let us stop here and analyze the results.

Areas of polygons up to 20 sides are calculated by our method and were exactly matching with the areas available on wiki.

It confirms that our method for calculation is correct and moreover we have 20 new friends of π.

Let us calculate areas of polygons with more number of sides to explore the path where others have probably not gone. The next table will show results.

Table 3

Number of sides, n	R_n	π_n	P_n	A_n
21	3.317283a	3.165240	21a	34.831474a^2
22	3.477576a	3.163122	22a	38.253340a^2
23	3.637770a	3.161277	23a	41.834356a^2
24	3.797877a	3.159659	24a	45.574524a^2
25	3.957907a	3.158234	25a	49.473844a^2
50	7.947272a	3.145733	50a	198.681810a^2
100	15.910258a	3.142626	100a	795.512898a^2
200	31.828370a	3.141851	200a	3182.837058a^2

Number of sides, n	R_n	π_n	P_n	A_n
300	47.744737a	3.141707	300a	7161.710637a^2
400	63.660668a	3.141657	400a	12732.133646a^2
500	79.576424a	3.141634	500a	19894.106086a^2
1000	159.154419a	3.141603	1000a	79577.209746a^2
2000	318.309624a	3.141595	2000a	318309.624384a^2
5000	795.774610a	3.141593	5000a	1989436.526849a^2
10000	1591.549378a	3.141592	10000a	7957746.892795a^2
20000	3183.098835a	3.141592	20000a	31830988.356579a^2

Practically we are in a position to know π_n and area of any regular polygon.

We can conclude following things for any regular polygon:

'Ratio between the perimeter of the polygon and twice the Apothem is a constant (π_n).'

And

'Value of π_n starts from 5.1965 (for triangle) and approaches the value of 3.141592 (or π of circle) as the number of side increases.'

CHAPTER ELEVEN

End of the journey

Are you thinking, what will happen if 'n' is keep on increasing? It is an obvious question for all reader. Let's find out what happens when n becomes infinite (∞).In Mathematical sense, ∞ is something which is too large to achieve.

So, what happens to R_n and π_n?

We know that

$$R_n = \frac{a}{2tan\frac{180°}{n}}$$

At n = ∞

$$R_n = \frac{a}{2tan\frac{180°}{\infty}}$$

Practically you cannot explain $^1/_\infty$.

So, R_n = undefined at n = ∞

And for the same reason π_n is also undefined at n = ∞.

So, is this the end of our journey? Probably not!

Understanding R_n at n= ∞

To understand the situation Let us observe how R_n is changing with increasing number of sides of polygon (n).

In the above formula for R_n, if the side of the polygon, ‘a’ is fixed and number of sides is varied then the following table will emerge. Here θ is the angle made by one side of polygon on the center of polygon. The picture shown for each polygon is to show the variation of θ and apothem with varying number of sides of polygon.

Table 4

n	$\theta = 360^o/n$	$\theta = \angle BOC$
3	120°	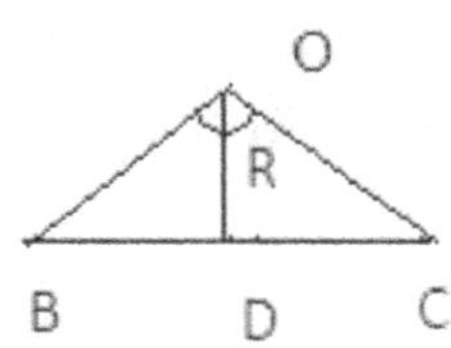
4	90^0	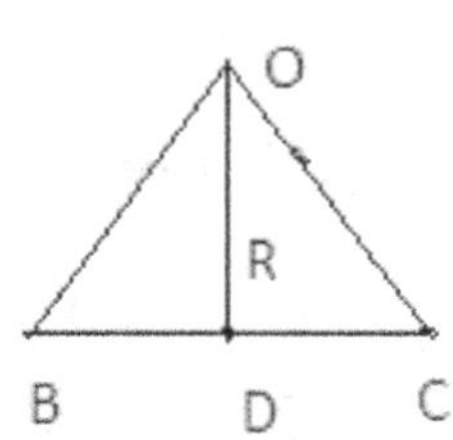

n	$\theta = 360^o/n$	$\theta = \angle BOC$
5	72^0	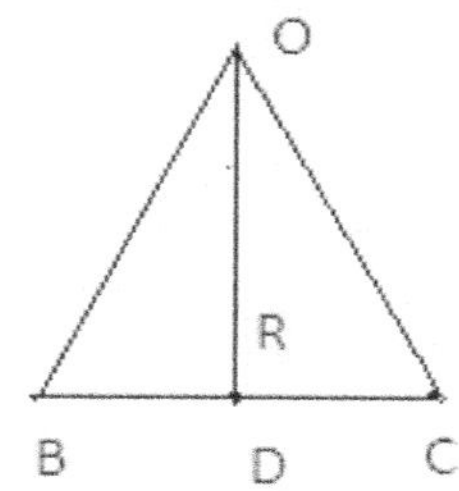
6	60^0	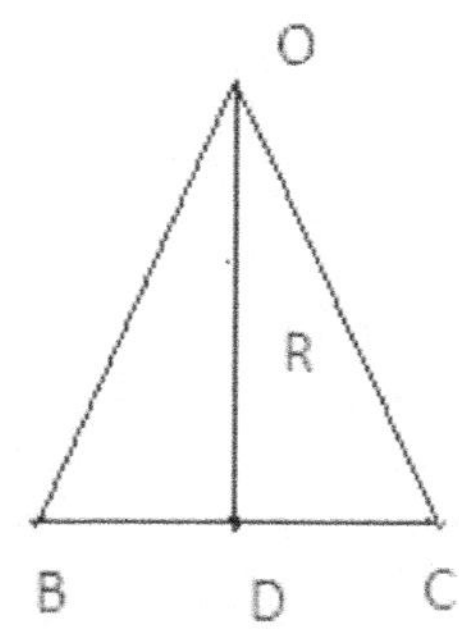
10	36^0	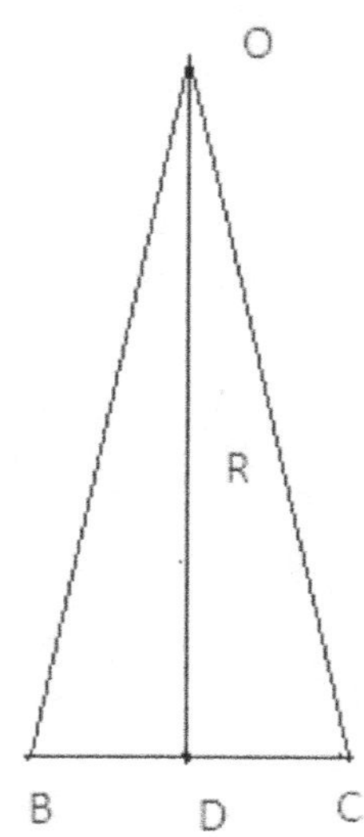

With the above table we can see that, for the same side BC = a, as the number of sides of polygon is increasing the value of Θ is decreasing and also the Apothem or R_n is increasing. So for a very large n, the size of R_n will be too large and can be treated as infinity.

It can also be seen that as the sides of polygon, n, is increasing the length of OB or OC (distance from center to corner) is also increasing and getting closer to Apothem OD ($=R_n$).

Applying **Pythagoras' theorem** in ΔOBD

$$OB^2=BD^2+OD^2,$$

$$OB^2 = \left(\frac{a}{2}\right)^2 + R_n^2,$$

$$OB=\frac{a}{2}\sqrt{1+\frac{1}{\tan^2\frac{180°}{n}}}$$

Above is the relation between OB and number of side of polygon. In the following table the change in R_n and OB with changing n is shown:

Table 5

n	R_n	OB
3	0.288675a	0.381881a
4	0.5a	0.559017a
5	0.688191a	0.732193a
10	1.538841a	1.559017a
20	3.156875a	3.166759a
30	4.757182a	4.763746a
50	7.947272a	7.951203a
100	15.910258a	15.912222a
200	31.828370a	31.829352a
300	47.744737a	47.745392a
500	79.576424a	79.576817a
1000	159.154419a	159.154615a
5000	795.774610a	795.774650a
10000	1591.549378a	1591.549398a

It can be easily observed from the above table that as the number of sides of polygon increases the difference between R_n and OB is reducing and beyond n=100 it is almost equal up to two decimal digits, above n= 500 it is three decimal digits and above n=5000 its 4 decimal digits.

R_n is the smallest distance from center to any side and OB is the largest distance from center. For very large values of n, R_n and OB are same which means that all points on the perimeter of such polygon are at equal distance from center.

Mathematical Solution for n=∞

In ΔOBD

Angle BOD = $^{\theta}/_{2}$,

And cos $^{\theta}/_{2}$ = $^{OD}/_{OB}$

As the value of n approaches ∞, the value of θ approaches closer to zero and correspondingly value of cos $^{\theta}/_{2}$ approaches 1.

So, as the value of n nears ∞, $^{OD}/_{OB}$ approaches 1.

Or we can say, for very large values of n, OD=OB.

And thus we can conclude that

'For a polygon with ∞ sides, the distance from center to any point on polygon can be treated as its apothem or R_n'

Effect of 'a' on R_n for larger values of n

As we know, if 'a' is the size of side of polygon then

$$R_n = \frac{a}{2\tan\frac{180°}{n}}$$

Above equation implies that value of R_n depends upon 'a'. The smaller 'a' is, the smaller will be R_n.

Let's understand this with an Example:

For a regular polygon having a=1 mm and n=50

R_n =7.95 mm

Have you ever imagined, with your sharp pencil, having tip size of 0.3 mm, how many dots are required to make a circle of 8 mm radius? And can you equate it with an 'n' sided polygon? Try it, it's interesting!

Understanding π_n at n= ∞

$$\pi_n = n\,tan\,\frac{180°}{n}$$

At n = ∞

$$\pi_\infty = \mathrm{n}\tan\frac{180°}{\infty}$$

As $^1/_\infty$ is undefined, so does π_∞.

But we have seen that the value of π_n starts from 5.19615 (for n = 3) and approaches towards 3.141592 (for n = 20000). If we put the above formula in MS excel and keep increasing n then, at n = 71231031 the calculated value of π_n is found to be 3.14159265358979 (which is equal to mathematical constant π for up to 15 digits). MS excel on my PC has a limitation; it can display only 15 significant digits. So for any value more than n = 71231031 excel will result only above value of π_n.

Following graphs emerges when π_n is plotted for different values of n.

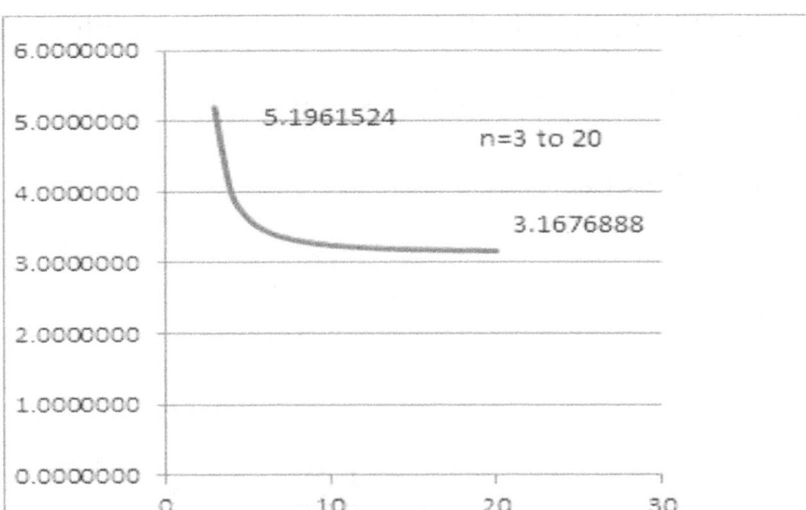

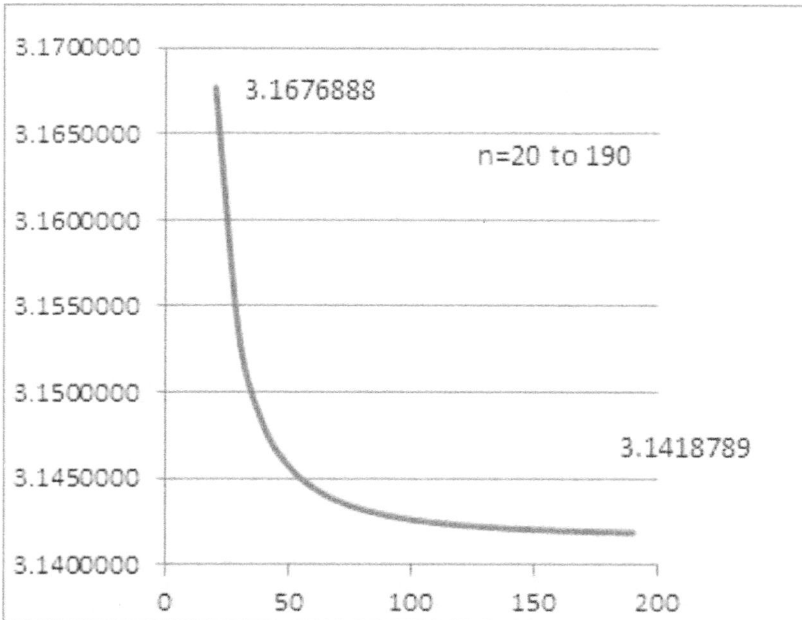
3.1700000
3.1650000
3.1600000
3.1550000
3.1500000
3.1450000
3.1400000
3.1676888
n=20 to 190
3.1418789
0
50
100
150
200

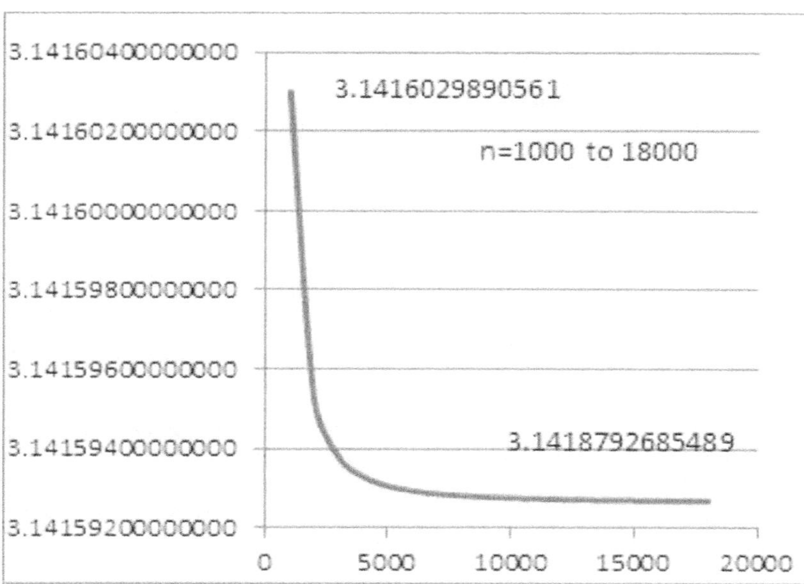
3.141604000000000
3.141602000000000
3.141600000000000
3.141598000000000
3.141596000000000
3.141594000000000
3.141592000000000
3.1416029890561
n=1000 to 18000
3.1418792685489
0
5000
10000
15000
20000

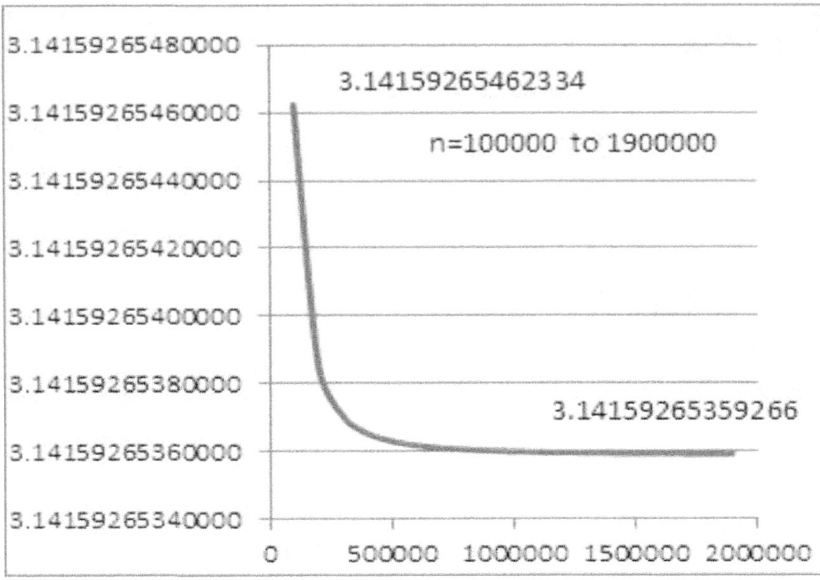
3.14159265480000
3.14159265460000
3.14159265440000
3.14159265420000
3.14159265400000
3.14159265380000
3.14159265360000
3.14159265340000
3.14159265462334
n=100000 to 1900000
3.14159265359266
0
500000
1000000
1500000
2000000

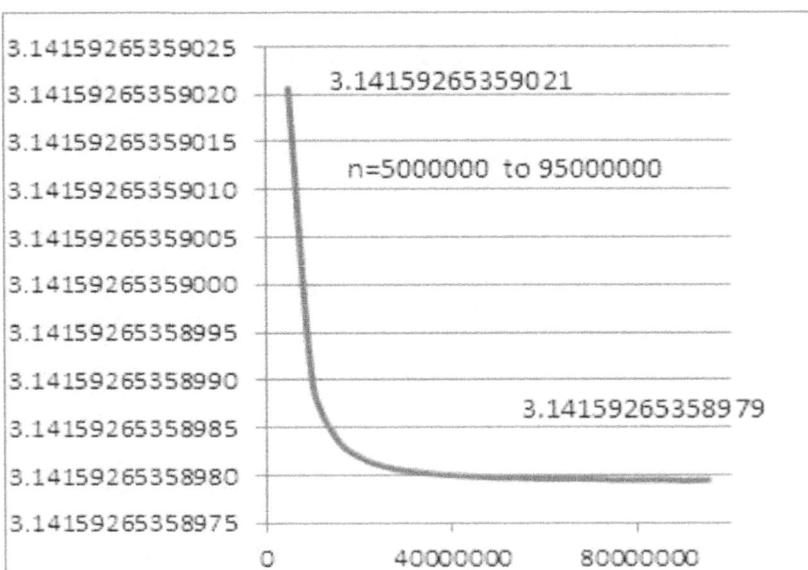
3.14159265359025
3.14159265359020
3.14159265359015
3.14159265359010
3.14159265359005
3.14159265359000
3.14159265358995
3.14159265358990
3.14159265358985
3.14159265358980
3.14159265358975
3.14159265359021
n=5000000 to 95000000
3.14159265358979
0
40000000
80000000

We can see that as 'n' is increasing the value of π_n is reaching towards a constant value whose significant digits are keeping on increasing as we take larger values of n.

Thus we can conclude that as 'n' tends towards ∞ the value of π_n approaches the mathematical constant π.

At $n = \infty$, $\pi_\infty = \pi$

$$\pi = \lim_{n\to\infty}\left(n \tan\frac{180^\circ}{n}\right)$$

Thus we can say that for a polygon with $n = \infty$:

a) *'R_n can be replaced by distance between any point on polygon and its center or say Radius of circle'.*

b) $R_\infty = R$ *(Apothem of polygon with ∞ sides OR radius),*

c) $\pi_\infty = \pi$ *(Mathematical constant or Archimedes Constant),*

d) $P_\infty = 2\pi R$ *(Perimeter),*

e) $A_\infty = \pi R^2$ *(Area)*

We can see that the area and perimeter of ∞ sided polygon is the same as that of a circle.

It seems that a circle is also a polygon with infinite sides.

CHAPTER TWELVE

Accuracy of π

We all know that world records have been made and broken for accuracy of $\boldsymbol{\pi}$.Accuracy level of up to trillion or more digits has been achieved and people are still striving for more accurate digits. They need more accurate π to have more accurate values of perimeter and area of circle.

Let's draw a circle of 5 cm (centimeter) or 50 mm (millimeter) radius.

We will calculate the area and perimeter of this circle using π with different order of accuracy and see the difference.

Perimeter of circle using π with different order of accuracy

Table 6

Accuracy level of π (in digits)	π	Perimeter ($2\pi R$)	Error with respect to perimeter obtained through 15 digit accurate π	Error (in %)	Error with respect to 1000 km distance
15	3.14159265358979	314.159265358979	--	--	--
3	3.14	314	0.159265358978985	0.05	506.95m
4	3.141	314.1	0.059265358978962	0.02	188.64m
5	3.1415	314.15	0.009265358978951	$2.95x10^{-3}$	29.49m
6	3.14159	314.159	0.000265358978993	$8.45x10^{-5}$	84.46cm
7	3.141592	314.1592	0.000065358978986	$2.08x10^{-5}$	20.80cm
8	3.1415926	314.15926	0.000005358978967	$1.71x10^{-6}$	1.70cm
9	3.14159265	314.159265	0.000000358978980	$1.14x10^{-7}$	1.14mm
10	3.141592653	314.1592653	0.000000058978969	1.88x10-8	0.18mm

In the table, initially, perimeter of the circle is calculated using 15 digit accurate π. Subsequently perimeter of the same circle is calculated with 3 digit accurate π and the value is compared with the Perimeter obtained through 15 digit accurate π. Error and Percentage error is calculated. In the last column, to give a relative idea, the error is shown with respect to 1000 Km.

Similarly data is collected for different accuracy levels of π up to 10 digits.

With 5 digit accuracy, the error comes down to the order of 10^{-3} (error of 30 m in 1000 km).With 10 digit accuracy the error comes down to a level that we cannot measure it with a simple scale , less than 1 mm for 1000 km.

To have a sense of judgment let us compare it with the distance between Earth and the Moon (384,000 km). For such large distance the error of calculation with 10 digit accurate π will come out to be 72 mm – "length of your little finger".

Need more accuracy —

“Believe me π has no limit!”

CHAPTER THIRTEEN

Friends of π

Finally the natural laws prevailed and it seems π is no exception. We have seen that all the regular polygons exhibit the same properties which a circle does.

For all regular polygons of side 'a' and number of sides 'n':

The Apothem R_n can be obtained by:

$$R_n = \frac{a}{2\tan\frac{180^\circ}{n}}$$

When 'n' is ∞ or 'a' is very small, R_n can be treated as radius R (for a circle).

1) The ratio of Perimeter to twice the Apothem is constant (π_n) for each polygon.

2) Value of this constant, π_n, varies for individual polygon and can be obtained through

$$\pi_n = n\tan\frac{180°}{n}$$

Where n = number of side of polygon. Value of n can vary from 3 to ∞.

3) Perimeter of polygon can be obtained through

$P_n = 2\pi_n R_n$, or $P_n = na$

When n = ∞ or 'a' is very small, R_n can be treated as R.

4) Area of the polygon can be obtained through

$A_n = \pi_n R_n^{\ 2}$

When n = ∞ or 'a' is very small, R_n can be treated as R.

Cheer up! You don't need to remember formula of area for individual regular polygon.

So, what the new friends of π will do? Nobody knows them! Let's know what role they can play. Some simple problems may help us.

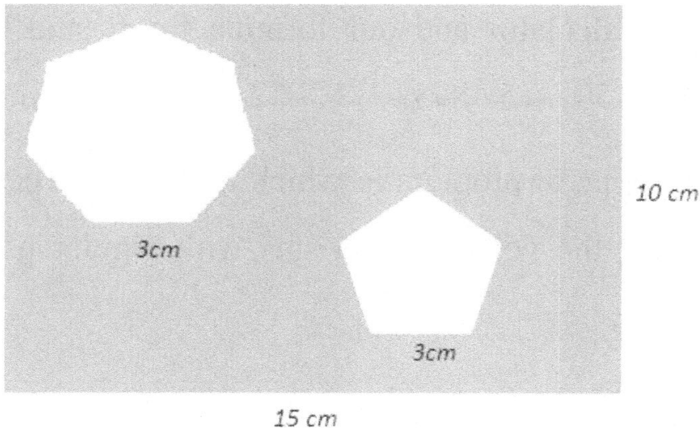

Above is a rectangular area of 15 cm x 10 cm. Inside the shape is a regular pentagon and a regular heptagon of 3 cm side . You have to calculate the area of grey shaded portion (area of rectangle without the two shapes).

Shaded area = $A_{rectangle} - A_{pentagon} - A_{heptagon}$

With conventional method of calculation it is quite a lengthy problem. But the new 'friends of Pi' can make it quite simple.

Area of regular polygon with side 'a'

$$A_n = \pi_n R_n^2 = \frac{n\,a^2}{4\tan\frac{180°}{n}}$$

Shaded area = $(15\text{x}10) - \pi_5 R_5^2 - \pi_7 R_7^2$

Using a calculator and our formula for π_n and R_n the answer is = 150 – 15.4843 – 32.7052 = 101.8105 cm^2.

Another type of problem in which friends of Pi can help is in finding the volume of prisms with regular polygon cross section.

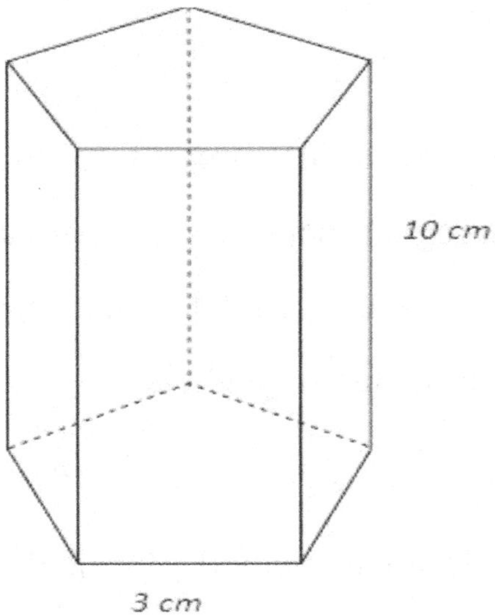

Above prism is having a regular pentagon cross section with 3cm (=a) side and 10 cm height (=h). Its volume can be easily calculated by using

Volume = $\pi_n R_n^2 h$

$$= \frac{n\,a^2}{4\tan\frac{180°}{n}}h = 154.84\ cm^3$$

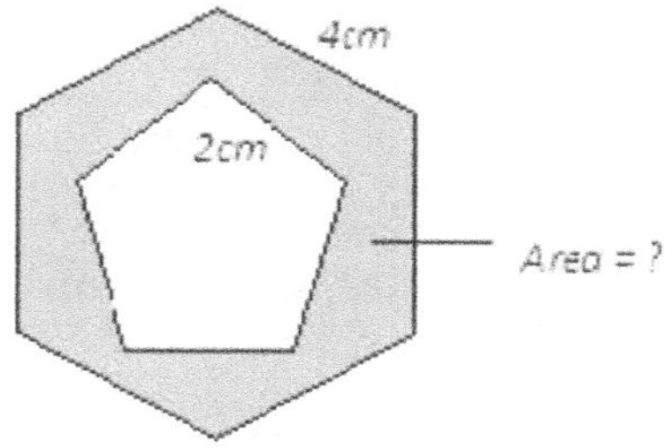

Another simple problem of calculating the shaded area between a regular hexagon (4cm) and a regular pentagon (2cm) is left for you. The calculation is difficult with known mathematics formulas but use of π_6 and π_5 makes it a child's play.

Above three problems were related to simple geometry. Friends of π can also help in finding the surface area of Regular Polyhedrons or **Platonic solids** (Tetrahedron, Cube, Octahedron, Dodecahedron, and Icosahedron). Its role in finding the volume of **Platonic solids** can be an interesting matter of research!

Apart from geometry, π plays important role in many areas of physics:

1) Approximate Time period T of a simple pendulum of length L

$$T \approx 2\pi \sqrt{\frac{L}{g}}$$

2) The Einstein's Gravitational constant $\boldsymbol{\kappa}$:

$$\kappa = \frac{8\pi G}{c^4}$$

'$\boldsymbol{G}$' is Gravitational constant and '$\boldsymbol{c}$' is speed of light.

3) Heisenberg's uncertainty principle:

$$\sigma_x \sigma_p \geq \frac{h}{4\pi}$$

Where $\boldsymbol{\sigma_x}$ and $\boldsymbol{\sigma_p}$ are standard deviation of position and momentum respectively, and '$\boldsymbol{h}$' is plank's constant.

4) Coulomb's law for the electric force

$$F = \frac{q_1 q_2}{4\pi \epsilon_0 \ r^2}$$

Where $\boldsymbol{q_1}$ and $\boldsymbol{q_2}$ are point charges, '$\boldsymbol{r}$' is the distance between, '$\boldsymbol{\varepsilon_0}$' is the electric constant, and $\boldsymbol{F}$ is the force between above two charges.

5) Magnetic permeability of free space:

$$\mu_0 = 4\pi . 10^{-7} \, N/A^2$$

6) Kepler's third law of planetary motion:

$$\frac{\alpha^3}{T^2} = \frac{G(M+m)}{4\pi^2}$$

a is semi- major axis of planet, ***T*** is the orbital period of planet around the Sun, ***G*** is gravitational constant ***M*** is mass of the Sun, ***m*** is mass of the planet.

7) The buckling formula for a long slender column under compressive stress is given by:

$$F = \frac{EI\pi^2}{(L)^2}$$

F is the critical load on the column; ***E*** is the modulus of elasticity, ***I*** is area moment of inertia for cross section of column, ***L*** is effective length of column.

It may be difficult for us to ignore '***friends of π***' in such equations. It can also be an interesting area of research!

Mathematics itself is full of equations consisting π. It is just a matter of time when it will embrace these **'friends of π'**.

So, the scope is endless!

Back to our search for friends of π – Now we can say that for every value of n (3 onwards) in the above equation for π_n, there exists a friend of π. One can name them as π_3, π_4, π_5, π_6 ……. endless.

We have seen that π has too many friends and we can say that—

"π is not alone!"

Author is a Mechanical engineer. He is working in Nuclear power plant in India.

You can contact him at social networking accounts:

Facebook - http://www.facebook.com/SujitSingh.npcil

Twitter - https://twitter.com/sujit_npcil

Made in the USA
Monee, IL
16 September 2026

63296735R00042